ISBN 978-1-333-93332-6
PIBN 10716000

This book is a reproduction of an important historical work. Forgotten Books uses
state-of-the-art technology to digitally reconstruct the work, preserving the original format
whilst repairing imperfections present in the aged copy. In rare cases, an imperfection in
the original, such as a blemish or missing page, may be replicated in our edition. We do,
however, repair the vast majority of imperfections successfully; any imperfections that
remain are intentionally left to preserve the state of such historical works.

English
Français
Deutsche
Italiano
Español
Português

# www.forgottenbooks.com

**Mythology** Photography **Fiction**
Fishing Christianity **Art** Cooking
Essays Buddhism Freemasonry
Medicine **Biology** Music **Ancient
Egypt** Evolution Carpentry Physics
Dance Geology **Mathematics** Fitness
Shakespeare **Folklore** Yoga Marketing
**Confidence** Immortality Biographies
Poetry **Psychology** Witchcraft
Electronics Chemistry History **Law**
Accounting **Philosophy** Anthropology
Alchemy Drama Quantum Mechanics
Atheism Sexual Health **Ancient History**
**Entrepreneurship** Languages Sport
Paleontology Needlework Islam
**Metaphysics** Investment Archaeology
Parenting Statistics Criminology
**Motivational**

# The Unilateral Dynamic Characteristics of Three-Electrode Vacuum Tubes.

BY

## JOHN G. FRAYNE

A THESIS

SUBMITTED TO THE FACULTY OF THE GRADUATE SCHOOL OF THE UNIVERSITY OF MINNESOTA IN PARTIAL FULFILLMENT OF THE REQUIREMENTS FOR THE DEGREE OF DOCTOR OF PHILOSOPHY.

Reprinted from the PHYSICAL REVIEW, N. S., Vol. XIX., No. 6, June, 1922.

[Reprinted from THE PHYSICAL REVIEW, N.S., Vol. XIX., No. 6, June, 1922.]

# THE UNILATERAL DYNAMIC CHARACTERISTICS OF THREE–ELECTRODE VACUUM TUBES.[1]

By JOHN G. FRAYNE.

### SYNOPSIS.

*Unilateral dynamic characteristics of vacuum tube when plate circuit includes resistance, inductance or capacity.*—(1) *Theoretical equations.* For the case of pure resistance ($R$), the Van der Bijl parabolic relation between plate current and effective grid voltage is expressed as a power series in $e \sin pt$, the impressed simple harmonic grid voltage. The *coefficients of the various harmonics* involve $R$, the $n$th harmonic reaching a maximum value when $R$ equals $(n-2)/3$ times $R_0$ the tube resistance. For the fundamental the *maximum energy output* for a given plate battery is secured when $R = 0.81 R_0$. The dynamic characteristic was obtained by compounding the harmonics into a single curve; it approaches a straight line as the resistance is increased. For the case of pure inductance, the plate current is expressed as a Fourier series. The dynamic characteristic is a closed loop whose area is proportional to the energy in the inductance. This loop reduces to an ellipse for small values of $e$, in which case the tube functions as an alternator whose internal impedance is a function of the external load. The insertion of a condenser instead of an equivalent inductance gives identical results except that the phase angle of the various harmonics is shifted. (2) *Experimental verification.* The effects on the plate current of varying the alternating grid voltage $e$, the static grid voltage $E_c$ and the plate voltage $E_b$, for a given value of resistance $R$ or inductance $l$, and the effect of varying $R$ or $l$ with constant $E_b$, $E_c$ and $e$ (15 or 20 volts), were determined and are shown in curves together with the corresponding theoretical values. A. W. E. 205B tube was used. The results show that the equations predict the harmonic constituents of the plate current as high as the fourth, for values of $e$ up to 15 or 20 volts (depending on $E_b$), the range for which the fundamental equation holds. For this range the coefficients of the various harmonics in the equation are proportional simply to $e^n$. The fundamental becomes greater while the other harmonics diminish as we approach the straight portion of the static characteristic and as we increase the plate potential.

*Circuit for producing pure sine wave electromotive force with frequency of 200,000 cycles.*—The oscillating circuit and filters used are shown diagrammatically in Fig. 1.

*Pure resistance for high frequencies.*—A platinized quartz fiber (diameter 0.01 mm.) with a resistance of 100 ohms per inch will carry 0.06 ampere when immersed in acid-free paraffin oil and has a negligible skin effect.

IT is a well-known fact that the current flowing from a hot filament to the plate of a three-electrode vacuum tube does not vary as the first power of the plate potential. With a view to determining what this relation really was, theoretical and experimental investigations were undertaken by Langmuir,[2] Bethenod,[3] Vallauri,[4] Van der Bijl,[5] Latour[6]

---

[1] Presented at the Chicago meeting of the American Physical Society, December, 1920.

[2] P. I. R. E., 3, 261–93, Sept., 1915, and PHYS. REV., 2, p. 457, 1913.

[3] La. Lum. El., 35, 25–31, Oct. 14, 1916.

[4] L'Elettrotecnica, Vol. N    8    d

and others. The second degree equation obtained by Van der Bijl lends itself more easily to mathematical treatment than any of the others, and agrees very closely with experimental evidence over a certain range of plate and grid potentials.

The curves obtained by plotting the plate current against the grid voltage for given plate potentials are usually referred to as the static characteristics of the tube. The term "dynamic characteristic" is used when the grid potential is of an oscillating nature. The latter characteristic is usually referred to as being "unilateral" when there is no external coupling between the grid and plate circuits, as distinguished from "regenerative" when such coupling exists. Van der Bijl has shown that the insertion of a resistance between the plate and the plate battery changes the form of the dynamic characteristic from a parabola to a curve which approaches a straight line with increasing resistance. A solution similar to that of Van der Bijl is obtained here, and in addition the case where the resistance is replaced by an inductance is worked out. We shall consider three cases here. First, with no resistance in the plate circuit, secondly, with a resistance inserted, and finally with the latter replaced by an inductance.

### CASE OF NO EXTERNAL RESISTANCE.

Let $E_b$ = plate potential,
$I_b$ = plate current,
$E_c$ = grid potential,

$e \sin pt$ = superimposed e.m.f. on grid.

According to the current-squared law

$$I_b = A(E_b + \mu E_c + \mu e \sin pt + \epsilon)^2, \qquad (1)$$

where $\mu$ is the amplification constant, defined by

$$\frac{1}{\mu} = -\left(\frac{dE_c}{dE_b}\right)_{I_b}$$

and $A$ and $\epsilon$ are constants depending on the structure of the tube. In this case the grid potential has the value $E_c + e \sin pt$. The equivalent plate potential is therefore $\mu (E_c + e \sin pt)$.

Before proceeding further it might be well to remark here that in order that (1) may actually represent the true conditions the value of $e$ must lie within certain limits, namely

$$e \leqq |E_c| + |g|, \quad e \leqq \left|\frac{E_b + \epsilon}{\mu}\right| - |E_c|,$$

where $g$ is the maximum positive voltage the grid can have before it begins to attract many electrons. Also if $e \sin pt$ attains such a large negative value in the cycle that the expression above is negative, the resulting current wave will be flattened out at that part of the characteristic curve. Equation (1) might be written generally as:

$$I_b = f(\mu e \sin pt).$$

Expanding by Maclaurin's Theorem

$$I_b = f(0) + \mu e \sin pt f'(0) + \frac{\mu^2 e^2 \sin^2 pt}{2!} f''(0)$$

and since $I_b$ is the same function of $E_b$ as it is of $\sin pt$

$$J'(0) = \left(\frac{d\,I_b}{d\,E_b}\right)_{t=0} = 2A(E_b + \mu E_c + \epsilon) = \frac{1}{R_0},$$

$$f''(0) = \left(\frac{d^2\,I_b}{d\,E_b{}^2}\right)_{t=0} = 2A.$$

Therefore

$$I_b = A(E_b + \mu E_c + \epsilon)^2 + \frac{\mu e \sin pt}{R} - \frac{A\mu^2 e^2}{2}\cos 2pt + \frac{A\mu^2 e^2}{2}. \quad (2a)$$

Thus in the simple case illustrated above where the plate potential is kept constant throughout the operation, a pure sine wave on the grid gives rise to a current of the same frequency (called the fundamental) in the plate circuit, a first harmonic and a rectified current component. In this case the actual dynamic and static characteristic curves will coincide.

### CASE OF A PURE RESISTANCE.

Next we shall consider the case where the potentials on the grid and plate vary simultaneously. Let a resistance $R$ be connected between the plate and the plate battery. Then

$$I_b = A\{E - RI_b + \mu(E_c + e \sin pt) + \epsilon\}^2 \quad (3)$$

This can be expanded as an infinite series.

$$I_b = f(0) + f'(0)\mu e \sin pt + \frac{f''(0)}{2!}$$

$$(\mu e \sin pt)^2 + \cdots \frac{f^n(0)}{n!}(\mu e \sin pt)^n \quad (4)$$

Van der Bijl has shown that since the parabolic relation connecting plate current and grid and plate potentials is only an empirical approximation, it is not to be expected that the higher derivatives in the series will accurately represent the actual experimental values. However, the derivatives up to probably the third or fourth ought to be a close approximation and the higher derivatives ought to indicate, at least in a qualitative way, how the higher harmonics depend on the various tube constants and on the properties of the external circuits.

Referring back to equation (4), the coefficients of the series are as follows:

$$f(0) = \frac{1}{2\,A R^2}\left\{\frac{1}{2}(B+1) - B^{1/2}\right\}, \quad B = 1 + \frac{2\,R}{R_0},$$

where

$R_0$ being defined as $\dfrac{1}{2\,A\,(E + \mu\,E_c + \epsilon)}$,

$$f'(0) = \frac{1}{R}\left\{1 - B^{-1/2}\right\}, \quad f''(0) = A\,B^{-3/2}, \quad f'''(0) = -2\,A^2 R\,B^{-5/2}$$

The general functional term is given by

$$f^n(0) = \frac{(-1)^n 2^{n-1} n(n-2)(n-4) - \cdots - R^{n-2} A^{n-1}}{n!\,B^{(2n-1)/2}}.$$

$$= \frac{(-1)^n 2^{n/2-1}\,R^{n-2}\,A^{n-1}}{\Gamma_{(n+1)/2}\,B^{(2n-1)/2}}$$

The coefficient of $\sin(pt)$ is therefore the value of this expression multiplied by $(\mu e)^n$. If this coefficient is denoted by $\alpha_n$, then

$$\left|\frac{\alpha_n + 1}{\alpha_n}\right| = \frac{2\,(n+2)\,R\,A\,\mu e}{(n+1)\,B},$$

$$\underset{n=\infty}{\text{Limit}} \left|\frac{\alpha_n + 1}{\alpha_n}\right| = \frac{2\,R\,A\,\mu e}{B}.$$

In order that the series (4) may be absolutely convergent

$$\frac{2\,R\,A\,\mu e}{B} < 1.$$

Therefore

$$e < \frac{B}{2\,RA\,\mu}, \qquad i.e., < \frac{1}{2\,A\mu}\left\{\frac{R + 2R}{2\,RR_0}\right\}.$$

Thus for a given $A$, $\mu$ and $R$, the smaller the value of $R_0$, the greater $e$ may be.

Using the values of $A$, $\mu$, $R$ and $R_0$ given later, $e$ may have values reaching up to 150 volts. However, it will be seen later that in practice $e$ cannot have a value larger than about 15 volts if equation (3) is to represent conditions accurately. The physical limitations which the tube imposes on the characteristic equation make it impossible to use grid voltages more than one tenth of the limiting value as given by (5). It is very evident that for small values of $e$, the series (4) converges rapidly and in consequence only a few terms need be evaluated in order to find a close approximation to the actual current flowing under a certain condition of the amplifier.

Now $f'(O)$ stands for the reciprocal of the total output resistance $R'_0$ when there is an external resistance in the plate circuit.

Therefore

$$\frac{1}{R'_0} = \frac{1}{R}\left\{1 - B^{-1/2}\right\}$$

The total resistance of the complete plate circuit is thus a rather complicated function of the external resistance and the internal output resistance of the tube when there was no resistance in the plate circuit.

Since the series (4) is a power series in sin $(pt)$ it is necessary to convert the various powers of sin $(pt)$ into first-power expressions of functions of multiples of $pt$, and expressions corresponding to the rectified currents. Since the series converges rapidly for values of input voltage within the limits (2), all powers of sin $(pt)$ beyond the fourth will be omitted.

$$I_b = \frac{1}{2\,A R^2}\left\{\frac{1}{2}(B+1) - B^{1/2}\right\}$$
$$+ \left[\left\{\frac{\mu e}{R}(1 - B^{-1/2})\right\} - \frac{3}{2}A^2 R\,\mu^3\,e^3\,B^{-5/2}\right]\sin(pt)$$
$$- \left[\frac{1}{2}A\,\mu^2 e^2\,B^{-3/2} + \frac{5}{2}A^3 R^2 \mu^4 e^4\,B^{-7/2}\right]\cos(2\,pt + \pi)$$
$$+ \frac{1}{2}\left[A^2\,R\,\mu^3\,e^3\,B^{-5/2}\right]\sin 3\,pt$$
$$+ \frac{5}{8}\left[A^3\,R^2\,\mu^4\,e^4\,B^{-7/2}\right]\cos 4\,pt + \cdots\cdots\cdots\cdots$$

Actual computation of the coefficients in this series show that for values of $e$ within the limits specified above, the series converges very rapidly.

For values of $e$ below 10 volts, actual computations show that the coefficient of sin $(pt)$ is practically a linear function of $e$; beyond ten volts the term involving $e$ becomes appreciable and the relation becomes more complex. Similarly the coefficient of cos $(2\,pt + \pi)$ varies as the square of $e$ up to about 10 volts. Since we have taken no powers higher than sin $pt$ the coefficients of the third and fourth harmonics vary directly as the cube and fourth powers respectively of $e$.

The relation between the coefficients and $R$, when the latter is variable, can best be shown by examining the condition for maxima. If we take the $n$th derivative as representing the coefficient of the general term, then the latter will be a maximum when

$$\frac{d}{dR}\left(f^n\,(0)\right) = 0,$$

*i.e.*, when

$$R = \frac{n-2}{3}\,R_0 \quad \text{or} \quad (1 + 2\,R/R_0)^{-1} = 0.$$

The latter equation has a solution, $R = \infty$. This is, obviously, the condition for a minimum. The first relation shows that $R$ must be a negative quantity for $n = 1$. Hence the fundamental has no real maximum. For $n = 2$, the maximum occurs at $R = 0$. For $n = 3$, the maximum occurs when the external resistance is one third of the tube resistance. For the higher derivatives, the position of the maxima occur at continuously increasing values of $R$.

If the amplitude of the impressed e.m.f. is less than 15 volts, equation (3) holds good for all values of $R$, when the plate voltage is maintained at 200 volts, and the grid voltage is $-7.5$. For values of $e$ over 15 volts, using the same grid and the plate potentials, equation (3) no longer holds. Hence the amplitudes of the harmonics as experimentally found for values of $e$ over 15 volts depend on other features of the amplifier. Since $E_c$ is $-7.5$ the grid will be raised to a positive potential of 7.5 volts during this cycle. In Fig. 5 it will be noticed that at this value of $E_c$ on the 200-volt parameter, the static characteristic begins to lose its parabolic nature and tends to flatten out. From the nature of the static characteristic it may be seen that the higher the plate voltage is raised the greater the values $e$ may have and remain within the proper limits. This amounts to saying that the smaller $R_0$ is, the greater the input voltage on the grid may be.

The dynamic characteristics for this case may be obtained as follows· The instantaneous values of the various harmonics for values of $pt$ between 0 and $2\pi$ are plotted, and then these constituent sine waves compounded to give the actual wave shape. If now the resulting periodic current is plotted along the $I_b$ axis and the input voltage plotted on the $E_c$ axis, the resulting curve will be the so-called dynamic characteristic of the tube under the specific conditions. It will be seen that if all terms but the fundamental had been neglected, the characteristic would have been a straight line. Addition, however, of the first harmonic causes the characteristic to have a definite curvature. The smaller the value of the external resistance, the more nearly does the curve approach the parabolic relation holding in the static case, and, of course, in the limiting case when $R$ is zero, the two characteristics coincide.

## CONDITION FOR MAXIMUM OUTPUT.

In connection with the expression for the internal resistance, it may be pointed out that the usual statement that the maximum power is obtained from a tube when the external resistance in the plate circuit is equal to the internal output resistance of the tube needs clarification. If by maximum power is meant the greatest power obtained from the fundamental frequency, the following is valid.

$$\text{Power} = R I^2 = \frac{\mu^2 e^2}{R} \left[ 1 - B^{-1/2} \right]^2. \tag{6}$$

Therefore

$$\frac{dP}{dR} = - \frac{\mu^2 e^2}{R} \left[ 1 - B^{-1/2} \right] \left[ \frac{1}{R} \left( 1 - B^{-1/2} \right) - \frac{2}{R} B^{-3/2} \right], \tag{7}$$

$$\frac{dP}{dR} = 0 \text{ for maximum } P.$$

Therefore

$$B^{3/2} = 2B - 1 \text{ or } \frac{R}{R_0} \cdot = \frac{1 \pm \sqrt{5}}{4} = 0.81. \tag{8}$$

The condition for a maximum dissipation of fundamental current energy is that the ratio of $R$ to the internal resistance when $R$ was zero is .81. This condition holds in the case where the maximum power is desired with a certain fixed-plate battery, and a variable resistance is available. The usual condition for maximum power, that the internal and external resistances be equal, is only true in this case if the actual plate potential is kept constant while the plate resistance is varied. The condition under which the above relation was obtained is the one most commonly met with in practice.

## CASE OF AN INDUCTANCE.

When an inductance, $l$, is placed between the plate and plate battery, the equation for the plate current may be written as follows:

$$I = A \left\{ E_b - l \frac{dI}{dt} + \mu (E_c + e \cos pt) + \epsilon \right\}^2 \quad \text{or} \tag{9}$$

$$B^2 - 2 B L \frac{dI}{dt} - 2LF \cos pt \frac{dI}{dt} + 2 BF \cos pt + L^2 \left( \frac{dI}{dt} \right)^2 + F^2 \cos^2 pt$$

where

$$B = A^{1/2} (E_b + \mu E_c + \epsilon). \quad L = A^{1/2}, \quad F = A^{1/2} \mu e. \tag{10}$$

A rigorous solution of this differential equation for $I$ is very difficult. but an approximate method of solving it may be legitimately utilized. Experimental evidence shows that $I$ is a rapidly converging Fourier series, and that the frequency of the fundamental is the same as the frequency of the input e.m.f. on the grid.

We can therefore write:

$$I = a_0/2 + \sum_{n=1}^{\infty} \alpha_n \sin npt + \sum_{n=1}^{\infty} \beta_n \cos npt \tag{11}$$

In terms of the exponential values for the sine and cosine,

$$2\alpha_n \sin npt + 2\beta_n \cos npt = (\beta_n - i\alpha_n) e^{inpt} + (\beta_n + i\alpha_n) e^{inpt}$$

Write

$$2a_n = \beta_n - i\alpha_n, \quad \text{and} \quad 2b_n = \beta_n + i\alpha_n. \tag{12}$$

Then

$$I = a_0/2 + \sum_{n=1}^{\infty} a_n e^{inpt} + \sum_{n=1}^{\infty} b_n e^{inpt}. \tag{13}$$

If we substitute this value of $I$ in equation (10), we can arrange the resulting terms in ascending orders of $e^{ipt}$ and in descending orders of $e^{ipt}$ The expression will not be given here as it is very lengthy and cumbersome. Since we have terms involving $e^{ipt}$ and $e^{ipt}$ and corresponding higher powers of $e$ on both sides of the equation, the coefficients of the like powers on either side may be equated. Hence we obtain a series of $2n$ equations from which the $a$'s and $b$'s can theoretically be determined. The general solution of these equations, while ideally possible, is impracticable without further assumptions as to the nature of the coefficients. We saw in the case of the resistance of the plate circuit, that only the first few terms of the series were of importance for values of $e$ within the limits of equation (2), and that for values of $e$ up to 10 or 15 volts, the amplitude of the fundamental varied approximately as the first power of $e$. If all the coefficients other than $a_0$, $a_1$, and $b_1$, are negligible, we have:

$$a_1 = \frac{BA^{1/2}\mu e}{1 + 2BLip}, \quad b_1 = \frac{BA^{1/2}\mu e}{1 - 2BLip}, \tag{14}$$

showing that, for this case, $a_1$ and $b_1$ are linear functions of $e$. Substituting the above values of $a_1$ and $b_1$ in the expressions for $a_2$ and $b_2$, we have:

$$a_2 = \frac{A\mu^2 e^2}{4 (1 + 2 BLip)^2 (1 + 4 BLip)} ;$$
$$b_2 = \frac{A\mu^2 e^2}{4 (1 - 2 BLip)^2 (1 - 4 BLip)} . \tag{15}$$

The values of $a_2$ and $b_2$ were found on the assumption that all the higher coefficients were negligible. Similarly $a_3$ and $b_3$ may be found and so on.

From relation (22) the values of $\alpha_1, \beta_1, \alpha_2, \beta_2$, etc., may be found, and

$$\alpha_1 \sin pt + \beta_1, \cos pt = \frac{\mu\, e \cos (pt - \alpha)}{(R_0 + l^2 p^2)^{1/2}} \text{ , where } \alpha = \tan^{-1}\frac{lp}{R_0} \quad (16)$$

$$\text{and } \dot{R}_0 = \frac{1}{2\, A\, (E_b + \mu E_c + \epsilon)} \text{ ,}$$

$$\alpha_2 \sin 2\, pt + \beta_2 \cos 2\, pt = \frac{A\mu^2 e^2 \cos (2\, pt - \beta)}{2\,(R_0^2 + l^2 p^2)\,(R_0^2 + 4\, l^2 p^2)^{1/2}} \text{ ,} \quad (17)$$

where

$$\beta = \tan^{-1}\frac{2lp}{R_0}\left[\frac{2\, R_0^2 - l^2 p^2}{R_0^2 - 5\, l^2 p^2}\right]$$

Similarly $\alpha_3 \sin 3\, pt + \beta_3 \cos 3\, pt$ may be found and so on for the higher terms.

Since $I = a_0/2 + \Sigma \alpha_n \sin (npt) + \Sigma \beta n \cos npt$, the addition of the various quantities found above will give the resulting current $I$.

It is obvious that as soon as the values of $a_2$ and $b_2$ become appreciable compared with $a_1$ and $b_1$, the values of the latter obtained above can no longer be correct, since they were determined on the basis that all the other coefficients were negligible.

If the values obtained for $a_2$ and $b_2$ are substituted in the equations for $a_1$ and $b_1$, the following is the value of

$$\alpha_1 \sin pt + \beta_1 \cos pt = \frac{\mu e \cos (pt\,(pt - \alpha)}{(R_0^2 + l^2 p^2)^{1/2}}$$

$$+ \frac{2\, R_0^5\, A^2\mu^3 e^3\, lp \cos (pt - \epsilon)}{(R_0^2 + l^2\, p^2)^2\,(R_0^2 + 4\, l^2 p^2)^{1/2}} \text{ ,} \quad (18)$$

where

$$\alpha = tan^{-1}\frac{lp}{R_0}\, \epsilon = tan^{-1}\frac{R_0}{2\, lp} \text{ and } \left(\frac{l^2 p^2 - 2\, R_0^2}{R_0^2 - 5\, l^2 p^2}\right).$$

In order to obtain a numerical value for $\alpha_1 \sin (pt) + \beta_1 \cos pt$, it is best to evaluate each term separately and then compound the results by the parallelogram law. Similarly if the values of $a_3$ and $b_3$ become comparable with $a_2$ and $b_2$, we find for the corrected value of

$$\alpha_2 \sin 2\, pt + \beta_2\cos 2\, pt = \frac{R_0^3\, A\mu^2 e^2 \cos (2\, pt - \beta)}{2\,(R_0^2 + l^2 p^2)\,(R_0^2 + 4 l^2 p^2)^{1/2}}$$

$$+ \frac{12\, R_0^7\, A^3\, \mu^4 e^4\, l_2 p_2 \cos (2\, pt - \lambda)}{(R_0^2 + l^2 p^2)^2\,(R_0^2 + 4\, l^2 p^2)\,(R_0^2 + 9 l^2 p^2)^{1/2}}$$

where $\beta$ is the same as defined in (27) and

$$\lambda = \tan^{-1}\frac{3lp}{R_0}\left(\frac{3\, R_0^4 - 17\, l^2 p^2 R_0^2 - 8\, l^4 p^4}{R_0^4 - 32\, l^2 p^2\, R_0^2 + 40\, l^4 p^4}\right).$$

By making successive approximations as many terms as desired may be included in the expressions for any particular harmonic. It will be noticed that the resulting angle of lag of each harmonic depends on the number of terms we include in the coefficient, and thus there arises a peculiarity in a vacuum-tube generator, namely that the angles of lag of the various output harmonics are dependent on the amplitude of the input wave on the grid. The larger the amplitude for the coefficients of the various harmonics, and the consequent shifting of the angles of lag results.

## AREA OF THE CHARACTERISTIC LOOP.

Since the fundamental plate current lags behind the grid voltage by an angle $\alpha = \tan^{-1}(lp/R)$, it is evident that if this current be plotted against the alternating grid voltage an elliptical characteristic will be produced. If, however, the first and higher harmonics are plotted in addition to the fundamental, and the curves thus formed compounded into a single curve, it is evident that the characteristic will no longer be a true ellipse. Since the amplitudes of the harmonics are small compared to that of the fundamental, the resulting curve will not seriously depart from an ellipse. This curve is what is usually referred to as the dynamic characteristic. The $a_0/2$ term of the series gives the point of operation on the static characteristic, and it is obvious from the expression for the latter, that the larger the harmonics become, the greater is the shift of this point of operation. In practice this shift is noticed by the increased reading of a direct-current milliameter

If all but the fundamental current is omitted, the area of the loop may be easily found.

Put $x = E_c + e \cos(pt)$

$$y = \frac{1}{2}a_0 + \frac{\mu e}{(R_0^2 + l^2 p^2)^{1/2}} \cos(pt - \alpha)$$

Limits for $pt$ are $o$ and $2\pi$

$$\text{Area} = \int_0^{2\pi} y\,dx = -\int_0^{2\pi} \left\{ a_0/2 + \frac{\mu e}{(R_0^2 + l^2 p^2)^{1/2}} \cos(pt - \alpha) \right\} \sin pt\, dt$$

$$= \frac{2\pi e^2 u \sin\alpha}{(R_0^2 + l^2 p^2)^{1/2}} = \frac{2\pi \mu e^2 lp}{(R_0^2 + l^2 p^2)}, \text{ since } \alpha = \tan^{-1}\frac{lp}{R_0}.$$

If the curve be referred to the $I_b$, $E_b$ axes, this expression must be multiplied by $\mu$. Also since the maximum value $I_0$ of the fundamental is $\mu e (R_0^2 + l^2 p^2)^{-1/2}$, the area of the loop may be written as

$$A = 2\pi I_0^2 lp = 2\pi \frac{lp}{R_0} R_0 I_0^2.$$

· Since $lI_0^2/2$ = the maximum dynamical energy in the inductance, the area of the loop is thus proportional to that quantity.    If the inductance were not present $A = 0$, which means that the characteristic no longer has the form of a loop, but reverts back to the type found when a resistance was placed in the plate circuit.

## CASE OF A CAPACITY IN THE PLATE CIRCUIT.

Since a condenser placed between the plate and the battery prevents the direct current from flowing, it is necessary to place a choke coil across the condenser.    The choke coil will allow the direct current to pass, but if made properly will offer almost an infinite resistance to the high·frequency current.

The solution for this case is directly analogous to that for the inductance problem, the only difference in the final result being that $1/cp$ always replaces $lp$.

Thus the simple expression for the fundamental becomes $\mu e/(R_0^2 + 1/c^2p^2)^{1/2} \cos(pt + \alpha)$ where $\alpha_1 = \tan^{-1} 1/R_0 cp$.    Similar expressions for the other harmonics may be found from comparison with the expressions found for the inductance.

The dynamic characteristic for this case is similar to that found for the inductance, the only difference being that it is traced out in the opposite direction.

## DESCRIPTION OF APPARATUS AND EXPERIMENTAL PROCEDURE.

In order to have an experimental set-up which could be used to verify the preceding theory, the following conditions and requirements had to be met.

(*a*) Production of a pure sine wave e.m.f.

(*b*) Use of a sufficiently low frequency that capacity effects might be of small magnitude.

(*c*) Accurate measurement of the input e.m.f. on the grid of the harmonic producing tube.

(*d*) Use of a pure resistance.

(*e*) Use of a pure inductance.

(*f*) Measurement of the amplitude of the harmonics produced, without introducing extraneous resistances, etc., into the harmonic producer.

Fig. 1 is the complete circuit diagram of the entire collection of apparatus used in the experimental work.    It may be divided into three main sections, the oscillator, harmonic producer and the harmonic analyzer.    The oscillator in the upper left corner is designed so as to produce as pure a sine wave as possible.    The tuned circuit $L^1C^1$ prevents the fundamental frequency from passing into the battery circuit, thus compelling it to travel to the filament through the inductance $L_3$ of the oscillating circuit. The condenser $C_1$ offers less and less impedance to

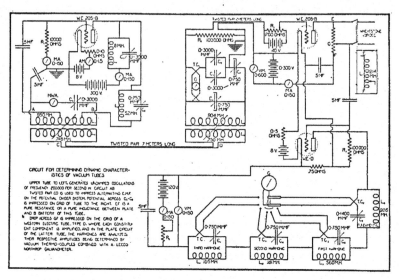

Fig. 1.

the higher harmonics, and the latter will pass down through $C_1$ to the filament terminal. As the first harmonic is always appreciable, it was specially filtered out of the oscillating circuit $L_3C_3$, by means of the anti-resonant circuit $L_2C_2$. The inductance $L_2 = .52$ M.H., of course offered some impedance to the fundamental frequency, but that was negligble compared with the impedance that $L_1C_1$ offered to the fundamental. These filters thus helped to produce a pure sine-wave oscillation of the same frequency as the fundamental in the circuit $L_3C_3$. The frequency used throughout was 200,000 cycles per sec., or a wave-length of 1,500 meters. This frequency was high enough that it could be tuned very sharply, and yet not so high that the internal capacities of the tubes would be of any importance. Ballantine[1] has worked out expressions for the input impedence of tubes under various conditions, and applying his formulæ to the W. E. 205 $B$ tube at this frequency and under the experimental conditions which will be described below, the input impedance was found of the order of 100,000 ohms.

The inductance $L_4$ was loosely coupled to $L_3$ and connected by means of a twisted pair with $L_5$, which in turn was loosely coupled to $L_6$. These latter coils were placed about seven meters away from the oscillating circuit, in order that they might not pick up any of the harmonics. The

---

[1] PHYS. REV., 15, p. 409–420, 1920.

loosening of the couplers already described resulted in maintenance of the sine e.m.f.  The combination of condensers $C_4$, $C_5$, $C_6$ and $C_7$ and the inductance $L_6$ is tuned for the fundamental frequency.  The arrangement of these condensers is what is known as a potential divider and has been described by Hulbert and Breit.[1]  The object is to take a portion of the alternating e.m.f. across $L_6$ and impress it on the grid of a tube. When the thermocouple is in the dotted position the current passing through $C_4$ is measured, but the current passing through $C_5$ and $C_6$ can easily be calculated when the values of the different capacities are known. The object of measuring the current in $C_4$ is that for small values of input potentials, the currents passing through $C_5$ and $C_6$ would be too small to be recorded by a low resistance thermo-couple.  If $I$ represents the amplitude of the alternating current passing through $C_5$ and $C_6$ then the resulting input e.m.f. is $I/2 \pi (C_5 + C_6) f$, where $f$ is the frequency of the wave.  For an e.m.f. of over one volt, the current $I$ could be measured directly by the low resistance thermo-couple.

The resistance $R_2$ is used to provide a leak for any charge that may accumulate on the grid, and allow it to flow to earth.  Its resistance must be comparable to the input impedance of the tube.  By means of potentiometer $R_3$ the potential on the grid could be varied as desired.

The upper tube to the right is the harmonic producer.  By means of the condenser potential divider a known value of input e.m.f. was impressed between the grid and filament and then according to equations (4), for a resistance $EF$ in the plate circuit, and (9) for an inductance $EF$, a plate current will result which is capable of being represented as a series of harmonics.  In order to get the results predicted in equation (4), $EF$ must be a pure resistance.  A straight wire immediately suggests itself as a resistance which would possess a minimum inductance and capacity.  However, in order to obtain a resistance of the order of 3,000 ohms, so much wire would be needed, that inductive and capacitive effects would become appreciable.  Then again, it is a well-known fact the conductivity of a wire diminishes with the frequency owing to the skin effect, and consequently the exact value of the resistance at any particular frequency is not easily determined.  A resistance suitable for high-frequency work should have a negligible skin effect, as well as having negligible inductance and capacity.  On the suggestion of Professor W. F. G. Swann, the author tried out some platinized quartz fibers immersed in acid-free paraffin oil, and found that they would carry currents up to at least 60 milliamperes.  From the formula for change in resistance with frequency,[2] it can be shown that using fibers about .01 mm.

[1] Phys. Rev., 4, p. 278, 1920.
[2] J. A. Fleming, Wireless Telegraphy, p. 97.

in diameter, the skin effect can be neglected. As the fibers used had a resistance of about 100 ohms per cm. only a short length of circuit was needed, thus reducing the inductance and capacity. The inductance *EF* was wound with No. 16, D.C.C. copper wire, and the windings were spaced about 1 mm. apart. The resistance of the coil was 10 ohms, whereas the reactance was 2,700 ohms.

In order to detect the various harmonics, a fraction of the e.m.f. along *EF* was impressed on the grid of a W. E. Co. V tube, this impressed e.m.f. being always less than 1 volt. A 100-ohm slide wire of IAIA wire was used for the variable portion of *EF*. It can be seen from equation (8) that in order to obtain pure amplification without the introduction of harmonics whose amplitudes are appreciable compared to that of the fundamental, the input e.m.f. must be small (less than one volt), and the value of the external resistance must be high. The resistance of an anti-resonant circuit is $R + (L^2\omega^2)/R$, where $L\omega$ is the inductive reactance. Since $R$, the ohmic resistance, is negligble, at radio frequencies, in comparison with $(L^2\omega^2)/R$, the latter term may be taken as the value of the resistance of an anti-resonant circuit. Therefore for any given $\omega$, $L$ should be made as large, and $R$ as small as possible. Now in the plate circuit of the tube which is used to separate out the harmonics, a series of anti-resonant circuits are placed. The first one is tuned for the fundamental, the second for the first harmonic, and so on. A vacuum thermocouple is placed in each circuit on the capacity side. This is done so that the D.C. plate current will not affect it. In order to keep the ohmic resistance low, thermocouples with heater resistances of from 0.5 ohm to 5 ohms were used, the higher resistance thermo-couples being used to measure the weaker amplitudes of the higher harmonics.

The effective resistance of these circuits are as follows: fundamental, 120,000 ohms; first harmonic, 183,000 ohms; second harmonic, 95,000 ohms; third harmonic, 175,000 ohms.

Arrangements (not shown in Fig. 1) were also made for measuring higher harmonics than these, by changing the inductance $L_{11}$, and by retuning $C_{11}$. Each thermo-couple could be connected successively to a Leeds and Northrup galvanometer, and the deflection of the latter indicated the root-mean square value of the alternating current passing through the heater. Previous to placing the thermo-couples in the circuits, they were calibrated using alternating current (60 cycles). It will be seen that the harmonic analyzer is essentially a voltage amplifier, picking out each frequency in the producer and magnifying its voltage. For this reason a tube with a large voltage amplification constant was chosen, in fact the value of $\mu$ as given by equation (4) was 26.

Let $I_1$ = the maximum current in the fundamental circuit.

Let $R_1$ = the effective resistance.

Let $R_0$ = the internal output resistance of the V tube.

Therefore $R_1 I_1$ = e.m.f. across $L_8$.

Let $e$ = e.m.f. across $FG$.

Let $\mu^1$ = actual voltage amplification factor.

Therefore $\mu^1 \quad \cdot \quad = \dfrac{\mu R_1}{R_0 + R_1} = \dfrac{26 \times 120,000}{29,250 + 120,000} = 20.8.$

Therefore $e \quad = \dfrac{R_1 I_1}{20.8} = \dfrac{120,000 \times I_1}{20.8}$

Let $r$ = resistance of $FG$.

Let $i$ = amplitude of current of fundamental frequency passing through $FG$.

Therefore $ri \quad = e$ and $i = \dfrac{120,000 \cdot I_1}{20.8 \times r}.$ $\qquad$ (15)

Thus knowing $I_1$ from the galvanometer deflection, and $r$ from the Wheatstone Bridge, the value of $i$ can be determined. For the case worked out above $i$ represents the amplitude of the fundamental fre-quency produced by a pure sine wave impressed on the grid of a tube having a pure resistance load in the plate circuit. Similarly, by measur-ing the currents in the other tuned circuits we can work back to the equivalent current in the harmonic producer.

When the resistance $EF$ is replaced by an inductance, a portion $FG$ of the inductance is used to obtain the input on the grid of the analyzer. In this case it will be noted that $EF$ offers twice as much impedance to the first harmonic, three times as much to the second harmonic, and so on. This makes it possible to measure weaker harmonics than in the case of the resistance. The inductance of $GF$ in this experiment was 0.0587 milli-henry, whereas the whole inductance of $EF$ was 2.14 milli-henries. The input impedance of the analyzer to which $EF$ was at-tached was of the order of 50,000 ohms at 200,000 cycles per sec., whereas the impedance of .0587 henry is only 74 ohms. This showst hat the impedance of the analyzer was practically short-circuited by the coil $FG$, and consequently did not affect the nature of the external circu it of the producer. For measurement of large output current values, the value of $FG$ was reduced to 0.04 M.H. To obtain, say the amplitude of the fundament current with the inductance, we have an equati on analogous to (15),

$$i = \frac{120,000\,I}{20.8\,(l\omega)},$$

where $l$ is the inductance of $FG$ and $\omega = 2\,\pi \times$ the frequency.

### EXPERIMENTAL RESULTS.

The following constants for the 205-*B* tube were determined from its static characteristic. $A = .554 \times 10^{-6}$, $\epsilon = 7.5$ volts, $\mu = 6.7$. For $E_b$ $= 260$ volts, $E_c = -7.5$ volts, $R_0 = 1/2\,A\,(E_b + \mu E_c + \epsilon)^* = 3,570$ ohms.

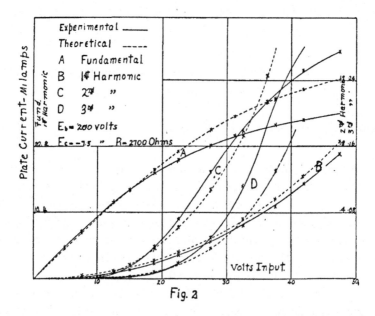

Fig. 2

*Resistance in Plate Circuit.*—When a resistance of 2,700 ohms was placed in series with the plate, and the value of $E_b$ reduced to 200 volts, the plate current was 21.5 milliamperes. Using these values of potential and resistance the curves shown in Fig. 2 were obtained. The amplitude of the input e.m.f. was varied from 0 up to 50 volts. The latter maximum loaded the tube rather heavily, and the larger currents were maintained just long enough to obtain the necessary readings.

Keeping the e.m.f. of the input at 15 volts, the actual plate potential at 200 volts, and varying the static grid potential the curves in Fig. 3 were obtained, for ranges of $E_c$ between $-30$ and $+12$ volts.

With a static voltage of $-7.5$ on the grid and the alternating e.m.f. kept at 15 volts, the variation of the harmonics with plate voltage was determined, as in Fig. 4.

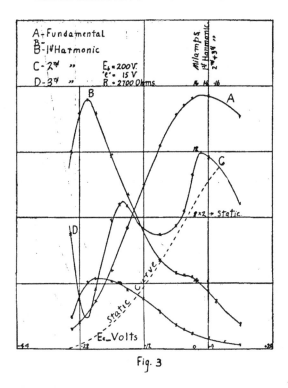

Fig. 3

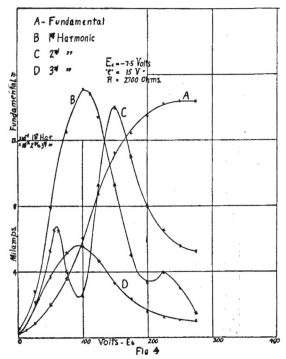

FIg. 4

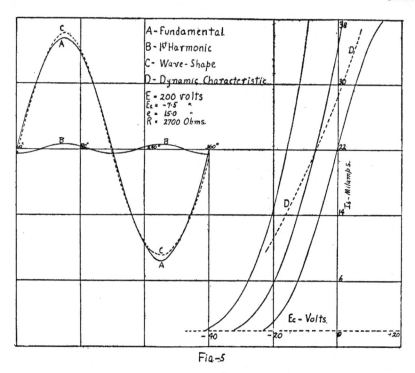

Fig. 5

Fig. 5 represents the wave-shape of the plate current obtained when a pure sine wave e.m.f. of 15 volts is impressed on the grid, the plate and grid potentials being the same as stated above. The fundamental and first harmonic are the only components included in the wave-shape. The

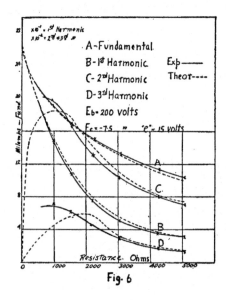

Fig. 6

amplitudes of the other harmonics are too small to be shown on the same scale. Fig. 5 also shows the dynamic characteristic for this case, where the wave-shape thus obtained is plotted against the alternating input voltage.

In Fig. 6 the variation of the harmonics with the value of the external resistance is shown.

*Inductance in Plate Circuit.*—Exactly similar experimental procedure was undertaken for the inductances. A higher plate voltage and plate current was used here, since there were no delicate platinized quartz fibers to be dealt with. For this case

$$E_b = 250 \text{ volts}, E_c = -10 \text{ volts}, R_0 = \frac{1}{2A\,(E_b + \mu E_c + \epsilon)} = 3{,}800 \text{ ohms}$$

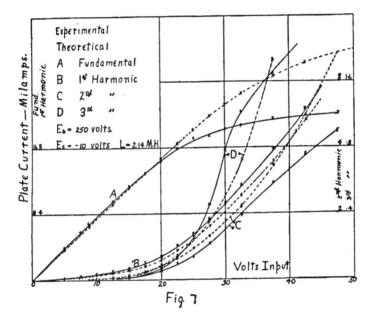

Fig. 7

The reactive load in the plate circuit was 2,700 ohms. Fig. 7 shows the variation of the harmonics with the alternating e.m.f. on the grid, under the conditions given above.

In Fig. 8 is shown the variation of the harmonics with the value of the static grid potential, the alternating input e.m.f. being constant at 20 volts, and the plate voltage being 250 volts. Fig. 9 shows how the variation of the static plate voltage affects the values of the harmonics.

The curves of Fig. 10 give the variation of the harmonics with the magnitude of the inductance in the plate circuit.

Fig. 11 represents the wave shape, using only the values of the fundamental and first harmonic.

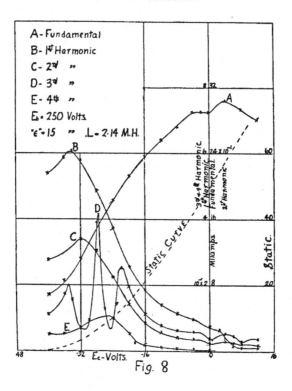

Fig. 8

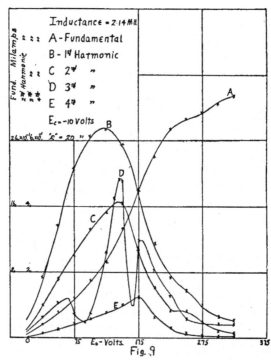

Fig. 9

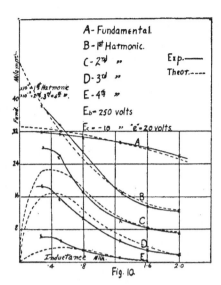

Fig. 10.

Fig. 11 also represents the dynamic characteristics for this case, the area of this loop being proportional to the amount of energy delivered.

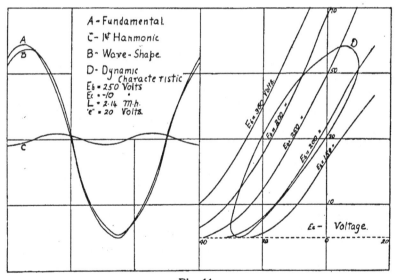

Fig. 11.

## DISCUSSION OF RESULTS.

The theoretical curves as shown in the various figures, are plotted from the values of the coefficients given by equations (3) and (8). In the case of the resistance the experimental fundamental values check up very closely with the theoretical values up to an input voltage of fifteen volts.

Beyond that voltage equation (3) no longer accurately represents conditions. It will be noticed that the maximum positive potential to which the grid is raised in this operation is 7.5 volts. Fig. 5 shows that at this voltage the static characteristic begins to flatten out, due to the passage of electrons to the grid instead of to the plate. For voltages higher than fifteen the theoretical first and third harmonics fall below the experimental values, whereas the second harmonic is greater than the experimental values would indicate. This would seem to be due to the flattening out of the wave at the upper end of the characteristic. In the case of the inductance good agreement between theoretical and experimental values were found for input voltages as high as 20 volts. This is due to the fact in this case that the operation was carried out over the 250-volt static characteristic, and it will be seen from Fig. 11 that this curve does not begin to flatten out until a positive grid voltage of ten volts is reached. This point of operation coincides with the maximum positive voltage to which the grid was raised with an input e.m.f. of 20 volts, $E_c$ being − 10 volts.

The curves of Figs. 3 and 8 show, in an emphatic manner, that the further we move towards the straighter portion of the static characteristic, the greater the fundamental becomes while the other harmonics continue to diminish. At − 40 volts the higher harmonics were still very much in evidence although the fundamental was rapidly approaching zero. The second harmonic in the resistance curves and the third in the inductance curves show rather peculiar irregularities. The sharp maxima and minima would at first seem to point to some sort of internal resonance in the tube. It will be noticed, however, that these are produced by simply varying either the grid or plate potentials, and are probably due to irregularities in the static characteristic which are smoothed out in the ordinary process of plotting. It will be recalled that in the case of the resistance the coefficient of the second harmonic is principally determined by the value of the third derivative of the current function. A small unnoticeable irregularity in the static characteristic might produce a large variation in the third derivative, and this effect is magnified by this method of analysis. In the case of the inductance the coefficients cannot be expressed as simple derivatives, but some such explanation as that given above will probably hold good here also.

The curves of Figs. 4 and 9 show that the greater the plate potential the greater the fundamental becomes while the harmonics continue to diminish. Beyond a certain plate potential, about 300 volts, the fundamental becomes nearly constant in value. The curves which exhibited the irregularities in the grid-variation series, also exhibit similar irregularities here. Further they occur at the same value of $(E_b + \mu E_c)$ showing that the effect is independent of whether the grid or plate potential is increased provided the sum as given is the same.

The curves of Figs. 6 and 10 show how the harmonics depend on the external impedance. Using this method of analysis it was impossible to use an impedance much less than 500 ohms. Even at this value, the input impedance of the analyzing tube cannot be neglected. Hence, beyond the first harmonic the experimental and theoretical curves do not agree very closely in this region. Although the experimental curves show signs of flattening out at this low impedance, they should have been rapidly approaching zero. For zero resistance or inductance, the fundamental and first harmonic have the values given by equation (2a), and all the other harmonics are zero, which is also in accordance with the same equation.

If all the harmonics were neglected the wave-shape for the resistance would be a pure sine wave in phase with the impressed e.m.f. These two, when compounded, would give a straight-line dynamic characteristic. It may be seen that with a sufficiently high resistance in the plate circuit this condition may be nearly reached. However, if the higher harmonic were also taken into consideration, the dynamic characteristic would be no longer linear but would have a curvature, which would be much less than that of the static curve. With the inductance, if all but the fundamental had been neglected the current would have been a pure sine wave lagging behind the impressed e.m.f. by an angle $\theta = \tan^{-1} lp/R$. The dynamic characteristic under those conditions would have been an ellipse. Addition of the other harmonic components tends to flatten out the sine wave, and consequently distort the purely elliptical characteristic.

In calculating the theoretical values of the coefficients of the various harmonics, up to a range of 15 or 20 volts, the coefficients of the fundamental were practically proportional to the first power of the applied e.m.f.; the first harmonic was proportional to the second power; and so on. It was only for values of the input voltage beyond 20 volts that the more complicated expressions for the coefficients had to be evaluated.

In conclusion, the writer wishes to express his very sincere thanks to the Western Electric Company, Inc., of New York, for their kindness in loaning necessary tubes and vacuum thermo-couples, and also to Professor W. F. G. Swann of this department for his many helpful suggestions and criticisms and his invaluable encouragement at all times.

DEPARTMENT OF PHYSICS,
    UNIVERSITY OF MINNESOTA,
        November 11, 1921.

CPSIA information can be obtained
at www.ICGtesting.com
Printed in the USA
BVOW06s1135270617

487931BV00011B/129/P